The Marshall Cavendish Library of Science Projects

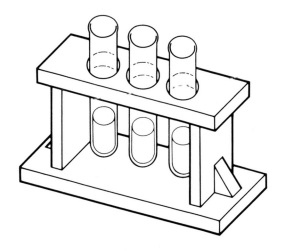

Marshall Cavendish Corporation

London • New York • Toronto

The Marshall Cavendish
SCIENCE PROJECT BOOK
of
LIGHT

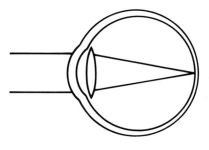

Written by Steve Parker
Illustrated by David Parr

Reference Edition published 1989

The Marshall Cavendish Library of Science projects
Light Volume 4

© Marshall Cavendish Limited MCMLXXXVI
© Templar Publishing Limited MCMLXXXV
Illustrations © Templar Publishing Limited MCMLXXXV

Trade edition published by Granada Publishing Limited

Reference edition published by:
 Marshall Cavendish Corporation
 147 West Merrick Road
 Freeport
 Long Island
 NY 11520

Printed and bound in Italy
by L.E.G.O. s.p.a., Vicenza

Library of Congress Cataloging in Publication Data

The Marshall Cavendish Library of Science Projects

 Includes index.
 Contents: (1) Water
 1. Science – Experiments – Juvenile Literature.
2. Science – Juvenile Literature. (1. Science –
Experiments. 2. Experiments) 1. Marshall Cavendish
Corporation.
Q164.M28   1986   507.8   86-11731
ISBN 0-86307-624-6 (Set)

ISBN 0-86307-628-9 (Vol 4 Light)

PICTURE CREDITS
Pages 6-7: Science Photo Library/A. Hart Davies
Page 8: Science Photo Library/David Parker
Pages 14-15: Tony Stone Photo Library, London;
right Science Photo Library/Robin Scagell
Pages 16-17: Tony Stone Photo Library, London
Page 21: Science Photo Library/Hank Morgan
Pages 22-23: Pictor International, London
Page 24: Science Photo Library/Orville/Andrews
Pages 26-27: Science Photo Library/David Parker
Pages 30-31: Tony Stone Photo Library, London
Page 37: Science Photo Library/Darwin Dale
Pages 40-41: Tony Stone Photo Library, London

CONTENTS

Science is all about discovering more about your world, finding out why certain things happen and how we can use them to help us in our everyday lives. SCIENCE PROJECTS looks at all these things. It's packed with exciting experiments and projects for you to do, and fascinating facts for you to remember. It will teach you more about the world around you and to understand how it works.

GOLDEN RULES

This book contains lots of scientific facts, experiments and projects to help you find out more about light and its strange properties. Whenever you try one of the experiments, make sure you read all about it before you start. You'll find a list of all the things you need, a step-by-step account of what to do, and finally an explanation of why and how your experiment works.

▶ Always watch what happens very carefully when you're doing an experiment and, if you find it doesn't work first time, *don't* give up. Consider what could have

gone wrong, and then read through the experiment once more. Check that everything is just right, and then try, try again. Real scientists may have to do an experiment several times before getting a worthwhile result.

▶ Because you will be such an active scientist, it's a good idea to start collecting for your laboratory. Nearly everything you need for the experiments can be found around your home. For example, bottles, cans and pieces of cardboard and paper will often be used, so when you see your parents throwing away useful containers,

■ GOOD SCIENTISTS...

ALWAYS THINK SAFETY FIRST

Famous scientists take precautions to avoid danger, so that they live to see their results and enjoy their fame. In any project or experiment, especially one you have thought up yourself, consider what it is you are trying to show and have a good idea of what should happen. Don't do any experiment without planning it 'just to see what happens'.

ALWAYS KEEP A NOTEBOOK

Whenever you are involved in scientific activity, keep your *Science Notebook* by your side and fill it with notes and sketches as you go along. Get into the habit of writing up your experiments and observations – your notes will come in handy in the future.

ALWAYS FOLLOW GOOD ADVICE

Advice and instructions, like the leaflets that come with pieces of equipment, should be read and understood. They are there for your safety and help. Good scientists think for themselves, but they are also wise and listen to what others have to say.

offer to wash them and then add them to your collection. General things like rulers, spoons for measuring, paper fasteners and a pair of scissors will also come in handy. You'll also need colored pens and paper for lots of the experiments, as well as sticky tape and glue. Last but not least, you'll need a work surface for your experiments and it's a good idea for this to be near a sink. Store your materials in a nearby cupboard or cardboard box.

▶ Always let your parents know what you are doing. Sometimes you'll need their help. And when it comes to special equipment like lenses or mirrors, they'll know where to get them. Your parents may also help you to build wooden stands or nail things down when needed. And if you need to use matches, cut things out, or drill holes, do remember to ask their permission first.

▶ Good scientists are clean and tidy! They always remember to clean up when they have finished! So after you have done your experiment or completed your project, throw away anything you won't need again and clean everything else, ready for next time.

NEVER MESS WITH THE MAIN LINE

Don't play with main line electricity or electrical outlets. Carry out all your electrical experiments with low-power batteries (preferably 3 or 4½ volts). Remember, main line electricity can kill you.

NEVER PLAY WITH CHEMICALS

Avoid mixing chemicals and powders unless you are sure that you know what is going to happen, and always use small quantities. Dangerous chemical mixtures can explode or start a fire or burn your eyes and skin. Make sure any chemicals you keep are properly stored in jars and are correctly labeled.

NEVER FOOL WITH HIGH-PRESSURE EQUIPMENT

Do not play around with gas or liquids under pressure, especially in containers like aerosol cans – even if they seem empty. They can blow up in your face. Dispose of empty aerosols carefully and *never* put them in or near a fire.

LIGHT OF THE WORLD

Have you ever been in a *really* dark place? So dark, that you couldn't even see your hand if you put it in front of your face? It can be quite scary, even though the dark is really nothing to be afraid of, and it's always good to get out into the light again.

Your sight – the way your eyes respond to light – is one of your most useful senses. Not just for finding your way around, or seeing where things are, but also for pleasure – looking at a beautiful view, for example or reading a book, or watching television or a movie.

All these things depend on light in one or more ways. What's more, understanding light and the way it behaves has also helped us to develop all sorts of interesting, sometimes life-saving, inventions – like microscopes, telescopes and cameras, lasers and mirrors for example.

Light has always had a powerful effect on our feelings and imagination. Long ago, when prehistoric people gathered around their night fires listening to the howling of the wolves and the screeching of unknown creatures in the dark forests, they must have welcomed the light of the fire as much as its warmth.

Today, of course, we don't have to rely on camp fires for light when the natural light of the Sun has disappeared for the night. We can have artificial light whenever we want it, just by flicking a switch. But on the whole we seldom think about light – except when it isn't there!

Have *you* ever wondered what light is, and why it behaves the way it does? Like so many other things, it's easy to

take light for granted. But it's not just convenient, or useful as a source of entertainment. As you'll find out in the following pages, light is extremely important – because all life on Earth depends on it.

7

WHAT IS LIGHT?

When you think of 'wavelengths', you probably think of radio waves – long wave, medium wave, short wave and so on. Radio waves are so called because radio signals actually do travel like waves. They spread out from the transmitter like ripples on a pond. The *wavelength* – 1500 yards, 247 yards, or whatever – is the distance between one wave and the next.

Radio waves can also be described by their *frequency,* that is, the number of wave peaks per second leaving the transmitter. Long wavelength signals have a low frequency, and short wavelengths a high one.

Radio waves are a form of what is called *electromagnetic radiation:* 'electromagnetic' because they are partly electrical and partly magnetic; and 'radiation' because they spread out and travel in all directions from their source.

So what's all this got to do with light, you may ask? Well, light is also a form of electromagnetic radiation, but it has a very much higher frequency (and shorter wavelength) than radio waves. It is made up of lots of energy particles called *photons*. There are lots of other types of 'wave', quite apart from radio waves and visible light. The 'electromagnetic spectrum' also includes microwaves, invisible infra-red and ultraviolet light, and x-rays.

White light?

The most important part of the electromagnetic spectrum to us is sunlight, which you can read more about on pages 14 and 15. It is often referred to as white light, but in fact sunlight is made up of seven different colors of visible light, plus infra-red and ultraviolet. These seven colors – red, orange, yellow, green,

What's a wave?

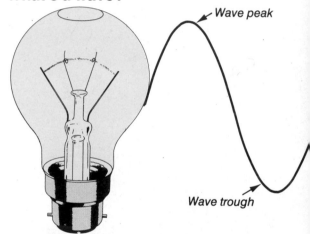

Light source

Wave peak

Wave trough

Science project

Seeing the spectrum

About 300 years ago a famous scientist called Sir Isaac Newton put a *prism* (a triangular piece of glass) in front of a beam of sunlight passing through a hole in a window blind. He found that the beam emerging from the other side of the prism had been split up into different colored rays. We now know that white light is made up of different colors of light, all traveling on different wavelengths. When they pass through the prism each one is bent by a different amount, resulting in the familiar 'fan' or spectrum of colors. The longest waves (red) are

White light enters a triangular prism and is split up to emerge as a spectrum of color.

8

blue, indigo and violet – normally travel as one in a straight line, moving from their source (the Sun) until they bounce or are *reflected* by an object. From there they travel in another straight line to our eyes. However, if those same light rays travel through another transparent medium apart from air – a glass of water, say – then they travel at a different speed and can bend or *refract* as they pass from one to the other.

When refraction takes place, the white light from the Sun may be split up to show its seven different colors, as you can see by doing the simple project below. You can read more about reflection and refraction on pages 22-27.

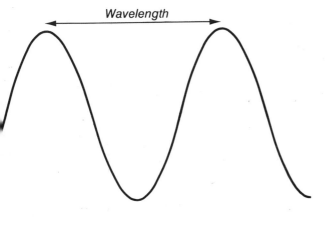

Wavelength

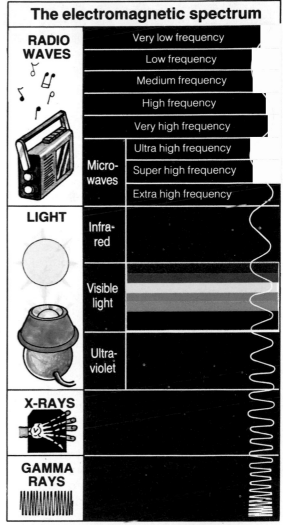

The electromagnetic spectrum

RADIO WAVES		Very low frequency
		Low frequency
		Medium frequency
		High frequency
		Very high frequency
	Micro-waves	Ultra high frequency
		Super high frequency
		Extra high frequency
LIGHT	Infra-red	
	Visible light	
	Ultra-violet	
X-RAYS		
GAMMA RAYS		

bent least and the shortest ones (violet) are bent most. This discovery was a great breakthrough in the understanding of light, but it took Newton a long time to convince people that his findings were true. Many of them laughed at his ideas at the time.

You can see the spectrum for yourself using just a glass of water, some cardboard and paper and, of course, some sunlight! First, cut a ⅜in (1 cm) wide slit in the cardboard and place it in front of a sunny window. Then place the jar on a sheet of white paper in front of the cardboard. The sunlight will pass through the water and split up to show a spectrum of color on the white paper. To work best, the sun needs to shine at an angle on the jar, so carry out this experiment in the morning or early evening when the sun is low in the sky.

Take a close look at the screen of a color television, and you'll see that all those bright, lifelike colors it produces are in fact made up of tiny stripes or dots of only three colors – red, green and blue.

These three colors are the primary colors of light that register in our eyes. By mixing them in various combinations and amounts, any color of the spectrum can be produced, and all three together in equal amounts produce white. If you take away the blue, for example, the red and green together give yellow. Subtract green and you get magenta, a purple-red color. Take away red and you get cyan, a greenish blue.

All the other colors that you see, like orange or brown or pink, are produced by mixing the primary colors in different proportions. If you take them all away (in other words, have no light at all) then you'll get black.

To anyone who's ever mixed up paints with a paint brush, this might sound rather strange. If you mixed up red, green and blue on a paint palette you'd get a sort of sludgy black color – certainly not white! This is because the primary colors of light and those of paint or *pigments* are quite different. Pigments only reflect light. They don't actually emit it in the way that the Sun or a TV set does, so they work in a quite different way. All the primaries of light mixed together will give you white, but if you do the same with pigments you'll get black.

How we see

Light enters your eye through the *pupil,* the circular opening at the front, and is focused by a tiny lens on to the light-sensitive *retina* at the back of the eye. The retina contains two types of cells: rods, which respond to the intensity of the light, and cones, which detect the colors. The rods and cones are connected to your brain by lots of nerves, which transmit the signals that your brain interprets as pictures.

The cones are sensitive to the red, green and blue wavelengths of light and all the different amounts that they might come in. But the rods don't provide the brain with much color information. Instead, they register the amount of light and are much more sensitive than the cones. They can

Primary colors

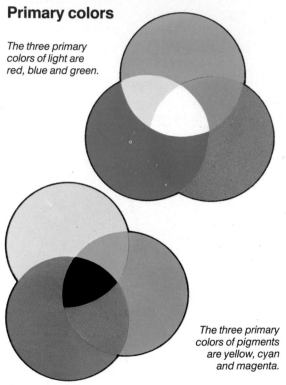

The three primary colors of light are red, blue and green.

The three primary colors of pigments are yellow, cyan and magenta.

Inside the eye

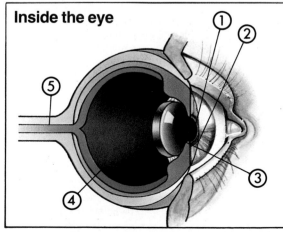

still operate in light which is much too dim for the cones to register any color, and this is why all colors look grey in dim light.

What is color?

Most of the objects we see do not produce their own light. Instead, they reflect light from other sources – either the Sun or an artificial source like a light bulb. But they don't always reflect the light shining on them in equal proportions and this is what gives them their color. A red rose, for instance, does not give out red light. Instead, its red pigment reflects the red part of the white light falling on it and this is what reaches your eyes. All the other colors are absorbed by the rose. Similarly, a yellow buttercup absorbs all the wavelengths of white light except the yellow which it reflects into your eyes.

How we see color

When we look at a red rose what we are seeing is the red part of the white light being reflected by the flower. All the other colors are absorbed. Similarly, with a yellow rose, only yellow is reflected.

1 Pupil – *A hole in the front of the eye which lets in light by enlarging in the dark and becoming smaller in bright light.*

2 Iris – *Controls the size of the pupil according to light conditions.*

3 Lens – *Bends the light coming in through the pupil and focuses it on the back of the eye.*

4 Retina – *Lining at the back of the eye which contains light-sensitive cells (rods and cones).*

5 Optic nerve – *Receives messages from the retina and carries them as nerve impulses to the brain.*

Science in action

Close-up on color

If you look at the color photographs or pictures in this book under a magnifying glass, you'll see that the colors are made up of tiny dots. These act just like the colored dots on a color TV screen to make up the picture that you see in front of you. All the different printed colors are made up of various combinations of just a few different colors of dots. These colors are the primary colors of pigments – magenta, cyan and yellow. The paper provides the white color and, in this book, black dots are also used to give better contrast. You can print color without using black dots by combining all the primary colors to make black, but the quality isn't so good.

When black-and-white pictures are printed, the image is made up of varying numbers of black dots. On newspaper pictures, these dots are often big enough to be seen with the naked eye.

You could try making up your own pictures by using just dots of red, blue and yellow paint. If you put lots of blue and yellow dots side by side, for example, that area of the paper will begin to look green.

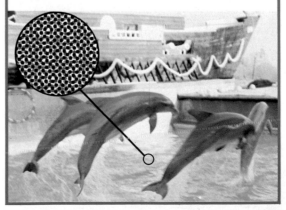

Your eyes, and the way they react to light and color, can sometimes play tricks on you. You can 'see' this for yourself by following the simple experiments shown here.

Experiment 1

Ghost colors

If you stare at an object for about twenty seconds – a candle flame, for instance – and then look away, you can often see a 'ghost' image of the object in front of you for a few seconds afterwards. Sometimes, this ghost image will be of a different color from the original. You can try this for yourself using some paper, some colored paints and a blank, white wall.

Step 1

Paint a large red cross on a big piece of white paper. Pin or stick it up on the blank wall and then stare at it for twenty seconds. This experiment will work best if you really concentrate on the cross and don't look away.

Step 2

Get a friend to quickly remove the piece of paper when the twenty seconds are up. Keep staring at the blank wall and you should see an after-image of the cross. But instead of being red like the original it will be a greenish-blue color – cyan.

Step 3

This happens because the red-sensitive parts of the cones in your retina have been registering the red of the cross for so long that they have temporarily stopped sending signals to your brain. The light bouncing back to your eyes from the blank white wall contains all the colors of the light spectrum which together make white. But when this hits your retina, the tired, red-sensitive receptors fail to react. The other receptors do respond, though, so the messages your brain receives make it see white with red subtracted, which is ... cyan!

You can produce different colored after-images simply by changing the color of the original cross. A green cross would result in a red 'ghost', a blue one would produce orange, a violet one yellow and so on. Try staring at the flag below for twenty seconds. You should be able to see a ghost image of a more familiar version on the wall!

Another trick which your eyes can play on you happens every time you watch a movie. The film is really a series of still pictures, each slightly different from the one before. Your eyes can only register up to 12 pictures per second as separate images, any more than this and you see them as a continuously changing single picture. This is why movie films run at a speed of 24 pictures (called frames) per second. Now try making your own simple movie by following the experiment on this page.

Experiment 2

Be a movie mogul

To make your own simple cartoon movie, all you need is a shallow metal tin (an old biscuit container would do), a strong, straight piece of bamboo, some paper, a piece of cardboard, some glue and some colored pens.

Step 1

Cut your paper to form a long strip which will wrap completely round the sides of the tin. Divide it up with faint pencil lines into twenty or so segments. In the first segment draw a picture of a monkey, or a clown or even a matchstick man. In the next segment draw the same figure but with its arms and legs in a slightly different position – as if it is walking along, for example. Repeat this process with the following segments until you have a series of pictures like those shown below.

Step 2

Get a grown-up to help you drill a hole in the center of the tin. This should be just big enough for you to wedge the bamboo stick into the hole. Push the stick through until about 4in (10 cms) is sticking out of the bottom of the tin. Sharpen this end into a point. Then glue your 'film' around the tin as shown.

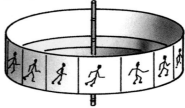

Step 3

Make a large pinhole in the center of your piece of cardboard and hold it close to one eye. Keep the other eye shut and then ask a friend to spin the tin round in front of you. You will need to position yourself so that the side of the tin is spinning directly in front of your eyes. Because the separate frames of your picture will be spinning round very fast, they will pass your eyes so quickly that they merge together to make one continuous, moving picture.

LIGHTING-UP TIME!

Look up at the sky on a clear winter night, and you will see a large number of the 6,000 or so separate stars that are visible from Earth with the naked eye. These are only a tiny fraction of the billions of stars in the galaxy, and they are all so far away that their light is too faint to be of any use to us.

The only star that really matters to us is our own Sun. As stars go, it's a pretty average example. It's not especially big or bright, but the important thing is that it's only about 93 million miles (149 million kilometers) away. This is fairly close in terms of space and near enough for it to provide us with energy as light and heat. You can read more about how the Sun produces this energy on pages 16 and 17.

Light travels through the vacuum of space at the speed of 186,282 miles per second (299,792 kms per second), making the speed of sound – a mere 1/5 mile (0.3 km) per second – a snail's pace by comparison. But even at this speed, the light from the Sun takes about eight minutes to reach us down here on the Earth.

The other stars that you can see are so far away that to describe their distance in terms of kilometers or miles means having to use awkwardly large numbers. For example, the nearest star visible to the naked eye (apart from our Sun) is *Alpha Centauri,* which is some 40 million million kms away (40 with twelve zeros after it!).

To make things simpler, astronomers measure such distances in *light years*. One light year is literally the distance that light travels in one year which is a mere 5.88 million million miles (9.46

million million kms). Using this system, Alpha Centauri is 4.29 light years away – a much easier number to handle than 40,000,000,000,000 kms!

Apart from the stars, the other main

feature of our night sky is the Moon. Unlike the Sun, the Moon produces no light of its own. Instead, what we call moonlight is really light from the Sun that is reflected down to Earth by the Moon. The same thing happens with other planets, such as Venus and Mars. They may look just like stars, but they too are really just giant reflectors of sunlight.

THE LIGHT OF LIFE

The light and heat we get from the Sun are very important to life on the Earth. But *how* important do you think they are? Without the Sun's warmth, which comes to us in the form of infra-red radiation, the Earth would freeze solid. But what would happen without the light? Would the loss of sunlight really be a big problem in these days of electricity and artificial light? After all, we manage all right at night! Surely we could do without sunlight if we had to. It would be boring perhaps, but not unbearable.

The ultimate provider

Unfortunately, the situation is not nearly so simple. Without light from the Sun all life on Earth would soon cease to exist. With the possible exception of some types of bacteria, all life on this planet depends ultimately on plants for food – either through eating them or eating the creatures that do feed on them. And in order to survive plants need ... light. Just as importantly, the life-giving oxygen in the air that we breathe is produced by plants. And in order to produce it plants need ... light! So you can see that without light we'd have no plants, and without plants we'd have no food and no oxygen to breathe.

Plant processors

The key to all this is a process called *photosynthesis*. During this process, plants take carbon dioxide from the air, and split it into carbon and oxygen. They use the carbon to build their cells and grow, and they release the oxygen back into the air for us to breathe. The energy they use to power these chemical reactions is the energy of sunlight. Take away the sunlight and the plants will die. And if the plants disappeared then so would we.

Plants need sunlight to live and grow. Without it, they could not power the chemical reactions that provide

Science in action

Energy from the Sun

The Sun produces a constant supply of light and heat for us down here on Earth. Both these things are forms of energy, but have you ever wondered where the Sun gets all this energy from? The answer is by burning hydrogen. Of course, it doesn't burn it in the same sense that we burn coal or gas. Instead, it uses it up in a nuclear reaction.

During this reaction, known as **fusion reaction**, four atoms of hydrogen are forced together to produce one atom of helium. At the same time, some of the hydrogen is converted into energy, in the form of light and heat. Thanks to this constant energy creation the Sun gets 'lighter (!)

us with oxygen to breathe and food to eat. Without sunlight plants would disappear, and so would we.

by about 4,600,000 tons every single second. Luckily for us the Sun has enough hydrogen left in it to last for many millions of years yet.

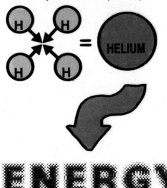

Four hydrogen atoms are forced together to produce one atom of helium and ... ENERGY.

Science project

No grow!

You can prove for yourself that plants need light in order to live and grow by doing this simple experiment. All you need is a plant with nice big leaves – like a geranium or a primrose – some black paper and some sticky tape.

Make sure your plant is healthy. Then choose one leaf and sandwich it carefully between two squares of the black paper. Tape round the edge so the leaf is securely covered. Leave the plant on a windowsill for a week and make sure it is kept moist. When you remove the paper you will find that the leaf underneath has become pale and unhealthy. This is because the paper prevented light from reaching the leaf and so starved it of the energy it needs to photosynthesize.

1 The leaf covered with black paper can no longer absorb light energy from the Sun. Without this energy, the leaf cannot power the chemical reactions that it needs to make its food.

2 After a week, the covered leaf will be pale and sickly. If it is left uncovered it will gradually recover.

LET THE ε ᴇ L ɢHT

Most animals start their day when the Sun comes up. At dusk the Sun sets and they settle down for the night. The Sun provides the light which begins and ends their day.

As usual, we humans don't like fitting in with nature. We have electric lights. At the flick of a switch we can banish darkness and fill a room – or even a football stadium – with bright, white light. Turning on a light is so much a part of everyday life that we never give it a single thought – unless there's a power cut, or the fuse blows, or the bulb goes.

Light makers

Throughout history people have used all sorts of light-making gadgets, so that far into the night and during the long winter evenings they could carry on with their lives. Primitive man had his camp fires and flaming wooden sticks, and in the Middle Ages wax candles and oily-rag torches flickered along the corridors of the great castles.

Around 1800 gas lights came into use. The gas arrived at your house by pipe

and was simply set on fire to give a wavering yellowish light. It wasn't very powerful, but is was better than candles for doing your homework!

The first electric lights came into use in the middle of the 19th century. They weren't like our bulbs of today, though. The light came from an electric spark jumping between the ends of two carbon rods. These lamps were called *carbon arc lamps*. They were quite bright but gave out a flickering light that could easily give you a headache, and you had to replace the carbon rods every few hours. So the search went on for better light-makers.

The first practical electrical light bulbs were invented by Joseph Swan in England, in 1878, and by the famous American scientist and inventor Thomas Edison, in 1879. Today's light bulbs work in exactly the same way as these early ones, as you can see in the diagram, although nowadays they are much more efficient and reliable. Even so, they always seem to go 'pop' at the most awkward moments.

How a fluorescent tube works

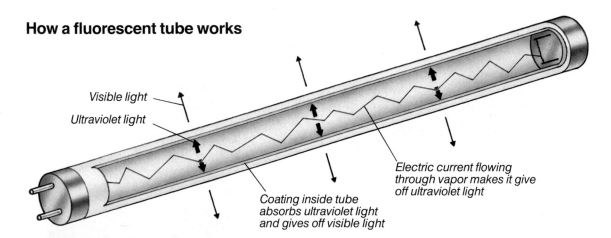

Visible light

Ultraviolet light

Electric current flowing through vapor makes it give off ultraviolet light

Coating inside tube absorbs ultraviolet light and gives off visible light

A common light in our homes, factories and offices is the fluorescent tube. This doesn't have a filament like the light bulb on the right. Instead, electricity flows through a gas, such as mercury vapor, inside the tube. The electricity makes the gas give

off light – but of the invisible, ultraviolet kind. This 'UV' light is then absorbed by a special chemical coating inside the tube and changed into normal white light. This is why the whole tube glows brightly.

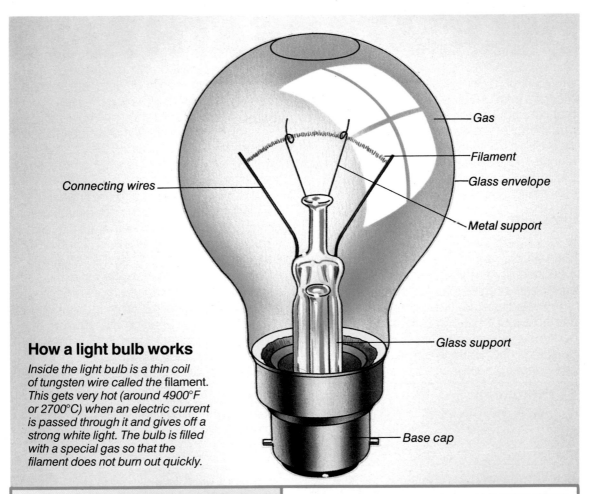

Gas

Filament

Glass envelope

Metal support

Connecting wires

Glass support

Base cap

How a light bulb works

Inside the light bulb is a thin coil of tungsten wire called the filament. *This gets very hot (around 4900°F or 2700°C) when an electric current is passed through it and gives off a strong white light. The bulb is filled with a special gas so that the filament does not burn out quickly.*

Science factfile

How bright is that light?

Many people need to measure light. Photographers, film-makers and TV-studio lighting crews, for example, would be lost without their light meters. Interior designers, too, know about lighting levels and how to change the look of a room by moving the lights and changing lampshades.

The main light-measuring unit is the *lumen*. This can be used in two ways. First, to measure the amount of light sent out by a source, such as the Sun or a light bulb. The second way is to measure the amount of light falling on a surface, such as the page of this book. This is measured in *lumens per square meter*. Lighting levels in offices and factories are measured in this way, to make sure that people have enough light to do certain jobs.

Light intensities

Source	Intensity in lumens
Sun at midday	160,000
Modern arc lamp	25,000
Electric filament bulb	500
Fluorescent tube	2
Candle	1
Moon (reflected from Sun	0.4

Light levels

Location	Lumens per square meter
Bright day outside	50,000 or more
Bright day with thin cloud	10,000
Indoors on sunny day	200
Minimum light for reading	50
Dark corridor or hallway	10
Moonlit night outside	0.1

DELICATE DEATH RAYS

The War of the Worlds is a science fiction story written by H G Wells. In it, invading Martians land on Earth, armed with weapons that fired not bullets, but death rays. Wells published this story in 1898, and from then on the death ray weapon was a great favorite of science fiction writers. It still is today, but the difference is that modern ideas are based on fact, not fiction.

Laser light

In 1960, a scientist called Theodore Maiman built a machine that could produce a beam of light more intense than had ever been seen before – a laser beam.

In the years since then, the laser has moved out of the laboratory and into everyday use – even into our own homes! Surgeons use lasers to carry out delicate eye operations, engineers use them to punch holes through thick solid steel, and even the latest record players use them to 'read' the signals on Compact Discs. They are used to create light shows at rock concerts, and 3-D images called holograms. And, of course, they can also be used as weapons. So what gives laser light its special abilities?

The answer is that, unlike ordinary light, laser light is more disciplined, as you can see from the diagrams below. Its wave peaks and troughs line up with each other which is what gives it its power. What's more, the waves stay together in a concentrated beam instead of spreading out in all directions like ordinary light waves.

The first lasers used small cylinders of ruby to produce their beams. In a ruby laser, the light energy from a powerful flash tube is absorbed by the atoms of the

Light waves

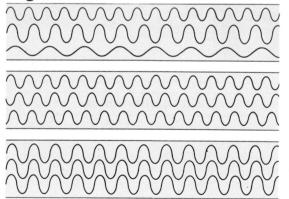

Ordinary light – a mixture of different wavelengths, all jumbled up.

Single color light – all waves of the same wavelength, but out of step with each other.

Laser light – all waves of the same wavelength, and all in step with each other.

A RUBY LASER

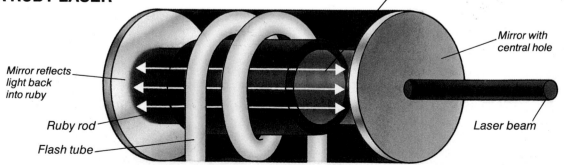

Mirror reflects light back into ruby

Ruby rod

Flash tube

Light emitted by ruby atoms bounces back and forth between mirrors

Mirror with central hole

Laser beam

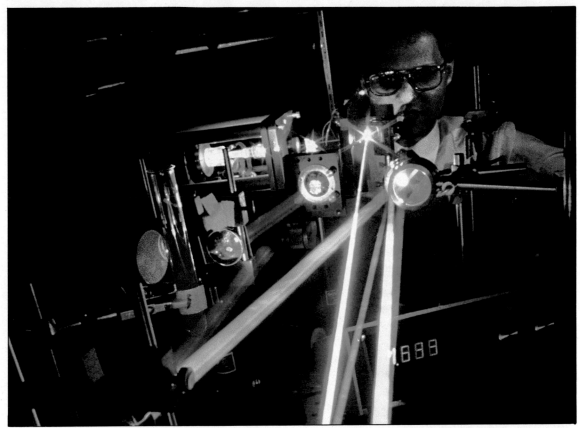

A technician puts a low-powered laser into action. This unique form of light is now used for a great many things – from micro-surgery to light shows at rock concerts.

ruby. But the atoms can only hold this energy for a short time. When they release it again, it is in the form of a burst of light particles which are reflected back and forth within the ruby, until they stream out of one end of it as a pulse of laser light.

Ruby lasers are still used, but they have been joined by a whole range of other types, based on different materials. These include gases, such as carbon dioxide, krypton and argon, and special liquid dyes. These all produce laser beams of different wavelengths and powers. It's because there are so many different types of laser that so many uses have been found for them. For example, low powered lasers are needed for surgery and light shows, but high powered ones are needed for cutting metals.

Measuring with lasers

Because laser beams don't spread out like ordinary light beams, they can travel enormous distances, and be aimed accurately. This has made them very useful for measuring great distances – and not just here on Earth.

In November 1970, the first unmanned Soviet Moon exploration vehicle, *Lunokhod 1,* landed in the moon's Sea of Rains. Amongst its instruments was a French-built laser reflector. When *Lunokhod* was in position, a beam from a ruby laser mounted in a telescope here on Earth was aimed at the reflector. By measuring the time it took for the beam to be reflected back to Earth again, scientists were able to make a very accurate calculation of the distance between the telescope and the reflector, and so check the distance from the Earth to the Moon (which is about 376,284 kms or 233,812 miles)!

I t's nearly midnight as Sam the super-scientist sets off in his car, to drive home along the dark country lanes. The gleam of the roadside reflectors and the white reflective lines painted on the road help him to follow the twists and turns of the lane. Suddenly the red reflectors of a parked car shine in front, and Sam has to pull out to pass it safely. Then, as he goes round a bend, a pair of yellow eyes shine out from the hedge – probably a farm cat out hunting, its eyes reflected in the glare of the headlights.

After looking in his rear-view mirror, Sam turns into his narrow driveway, guided by the orange reflector discs fixed to the gateposts on either side. At last he parks the car and goes into his house. Glancing at his reflection in the hall mirror, he notices how tired he looks. He must stop working so hard on this new formula for super-reflecting paint!

On reflection, you may find an awful lot of 'reflecting' happening on Sam's drive home from work. That's because there's always a lot of reflecting going on! Apart from the Sun and the light sources we make, such as electric lamps, few things create their own light. Instead, all you see are reflections – trees, houses, cars, animals and people all reflect light from a light source into your eyes. Of course, especially smooth surfaces like mirrors or glass, are the really good reflectors and show up best.

People have known about *reflection* since the earliest times. Prehistoric man probably noticed his face in the water as he knelt down at a pond to drink. He may also have noticed that stems and branches of nearby plants

seemed to bend where they went into the water, but when he pulled them out they were straight. We now know that this is due to *refraction,* or the 'bending' of light rays as they pass from one substance to another.

We can bounce light, bend it, twist it and focus it, all by applying the ideas behind reflection and refraction. You can read more about why light reflects and refracts, and how we can control it, on the following pages.

There are hundreds and hundreds of ways we use reflection and refraction in our lives. Remember this, next time you see yourself in the mirror, or look at a photograph, or put your spectacles on, or look down a microscope or...

ON REFLECTION

When you think of reflections, you may think of mirrors, or the smooth surface of a pond. You probably don't think of a carpet or a tree as being reflectors of light. But they are.

Since none of them produces light of their own (unless they're on fire), the only way you can see them is by the light they reflect. This means that any light rays hitting them are bounced back again, like a ball bouncing off a wall, and into your eyes. The difference between the reflection from a carpet and that from a mirror is this. The carpet has a very rough surface, so some of the light falling on it is absorbed. The rest is reflected – scattering in all directions by the bumps in the carpet. The mirror however has a smooth layer of silver behind the glass and this reflects nearly all the light straight back without scattering or absorbing it.

When you see yourself in the mirror what you're really seeing is light reflected onto the mirror by you now being bounced back again. You didn't make the light, though. It came originally from some other source – the Sun, perhaps, or a lamp. So it might go through several reflections before it reaches your eyes – off walls and doors, to you, to the mirror, and finally to your eyes!

Light colored surfaces are also much better at reflecting than dark ones. Go into a dark room with a piece of white paper and a flashlight. Shine the flashlight on the paper and then hold your hand about 1 foot (30 cms) above the paper. You should be able to see your hand quite clearly in the reflected light. Repeat the process using black paper. You will not be able to see your hand as well, since less light will be reflected onto it.

Here, the concentrated beam of an argon laser is bounced back and forth between four different mirrors.

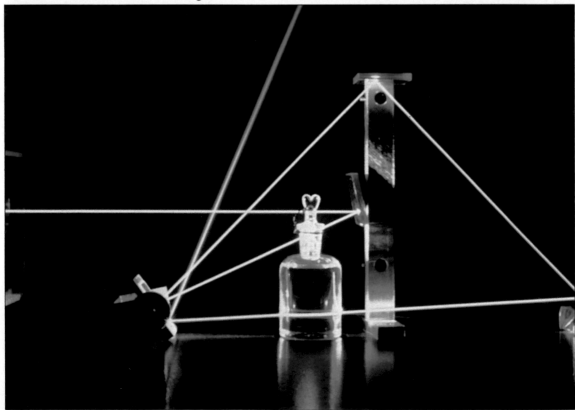

Image oddities

Next time you stand in front of a mirror, you might notice two odd things about your reflection. If you stand, say, a yard in front of a mirror, it's as if you were looking through the glass at someone standing a yard behind it – your image looks as though it's 2 yards away. This is because the image formed in the mirror is of yourself as 'seen' from a yard away. Add this to the distance between you and the mirror and you'll bring the total distance between you and your reflection to 2 yards.

You'll also notice that your image is the wrong way round. If you were facing another person instead of your own reflection, their right-hand side would be on your left, and vice versa. But your mirror image has your right-hand side on the right, because the light has been reflected back exactly the way it came.

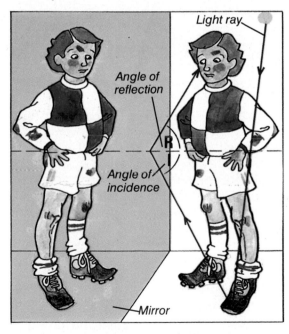

When a ray of light strikes a mirror, it bounces off it again at an equal angle. The angle at which the light hits the mirror is called the angle of incidence. *The angle at which it bounces off again is the* angle of reflection. *These two are measured from an imaginary line at right angles to the surface of the mirror, and are always equal.*

Science in action

Night eyes!

The 'cat's-eye' reflectors used in road markings were invented in 1934 by a road engineer called Percy Shaw. They are called cat's-eyes because they reflect light in much the same way as a cat's eyes does when light shines in them in the dark. A real cat's eye has a special membrane behind the retina (called the *tapetum*) which reflects light. This is what makes cat's eyes 'shine' in the dark – they are really reflecting any light that's around at the time, not producing light of their own as you might think. This also enables cats to see well in the dark, because the light not detected by the rods and cones of their retina on the way into the eye is reflected back through the retina and some is detected on this 'second pass'.

The 'eye' in the middle of the road has a similar design. It is made of a short glass cylinder, with a powerful lens at the front and a curved mirrored surface at the back. Light shining into it from a car's headlights passes through the lens and it focused on to the mirrored surface, which reflects it back out through the lens as a very bright reflection.

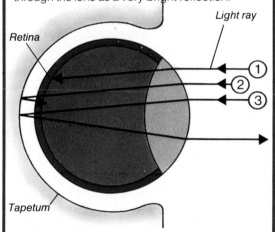

1   A ray of light is detected by the retina on its way in.
2   Another ray of light passes through the retina, is reflected by the tapotum and is finally detected by the retina on its way back out. (This is what gives a cat extra 'seeing power' in the dark.)
3   A third ray enters the eye as before but is not detected by the retina and passes back out of the eye, so making the cat's eye appear to 'shine' in the dark.

BEAM BENDING

Traveling through the nothingness of space, light moves at 186,282 miles (299,792 kms) per second. It moves more slowly through air. And when it travels through a much denser (but still transparent) material, such as glass or water, it moves more slowly still.

Refracting light

This slowing down of light has a very useful side-effect, for when light goes from water into the air it not only changes its speed, but also its direction – becoming slightly 'bent' at the junction between the two substances. This effect is known as *refraction* and when it was first discovered way back in the 17th century it opened up all kinds of interesting scientific possibilities.

Refraction is what makes a stick look bent at the point where it goes into the water. (You can see this for yourself just by sticking a pencil into a glass of water and watching what happens.) It is also what makes the bottoms of ponds and swimming pools look nearer than they are. You can see how this happens in the diagrams on the right.

Light and lenses

Understanding what happens when light passes from one substance to another was put to good use by scientists about 300 years ago. It was then that *lenses* were first made. These are specially-shaped pieces of glass which use refraction to make the image passing through them look bigger or smaller. Today, we use man-made lenses to help us with many things. You can read about some of them on the following pages.

One of the big problems that lens makers face is that different colors or wavelengths of light (the red wavelength,

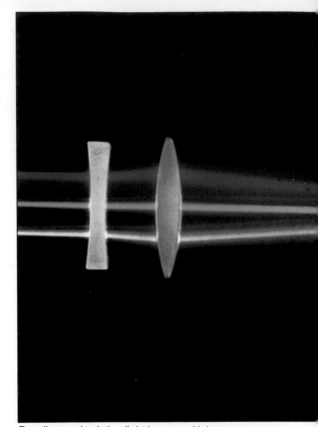

Bending and twisting light is easy with lenses; concave ones expand the beams, convex ones converge them and prisms change their direction altogether.

for example, or the green) are bent by different amounts, as you will know from the project on page 8. This effect is what causes a ray of sunlight to split up into a rainbow spectrum when passed through a glass prism. So with any lens there is a danger that the white light passing through it will start to split up into its component colors, each one being bent by a different amount. If this happens, the image seen through the lens will have colored fringes round it.

For this reason, high quality lenses, such as those used in cameras, are made of two or more different types of glass which are joined together by a transparent glue. Each type of glass bends the light by a slightly different amount so that color fringing can be avoided. You can read more about cameras and lenses on page 32.

Bottlebender

You can see refraction taking place very clearly by carrying out this simple project. All you need is a flat-sided glass bottle, some milk and water, a piece of cardboard and a flashlight.

Fill the bottle with water and add a few drops of milk. Lay this down in a dark room in front of your piece of cardboard in which you have cut a narrow slit. Shine the flashlight through the slit so a thin beam of light passes through the bottle. You will see how the beam bends as it meets the glass and again as it leaves it.

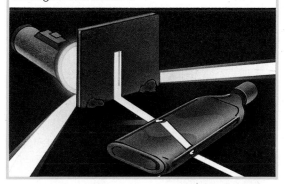

Why the bottom looks nearer!

Light reflected off the bottom of the pool is bent at the water's surface before it reaches the eye. Since we are used to light traveling in straight lines, our brains

don't take into account that the light from the bottom of the pool has been bent. So we see the bottom in a different place and nearer than it really is.

27

Have you ever caught sight of yourself reflected between two panes of glass or mirrors at right angles to each other? You may have noticed that your reflection looks different from usual. This is because your image is being reflected back and forth between the two mirrors to show the 'right way round', as others see you, rather than back to front as in your normal reflection. This is just one of the many mirror 'tricks' that you might come across. Here are some others for you to try.

Experiment 3

Mirror magic

To play these mirror tricks you will need a sheet of glass, a small light colored object, a glass jar, and a large and small mirror. For Step 3 you'll need a sheet of smoky perspex and the help of a friend.

Step 2

Hold your small mirror up in front of your larger one. Wiggle them about until you can see an endless line of mirrors in the larger mirror in front of you. Can you see how the image begins to fade as it gets smaller? This never-ending reflection happens because the light is being bounced back and forth from one mirror image to another, getting smaller each time.

Step 1

Place your small object about 6 inches (15 cms) in front of the sheet of glass. Position the glass jar behind the glass until it appears to be in the same place as the object's reflection (as in the drawing below). Now you have 'caught' the reflection you can measure its distance from the glass. You will find that it is exactly the same as the distance between the real object and the glass.

Step 3

Stand in front of the doorway into a room. Position your piece of smoky plexiglass just inside the room and angle it so that you can see the reflection of your friend standing inside and to the right of the door. Try putting a chair behind the plexiglass so you can see this too. With some practice you will be able to get your friend to 'walk' right through the chair.

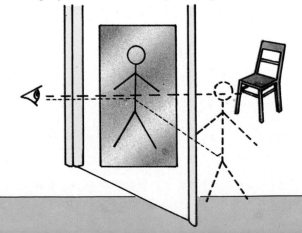

Experiment 4

Seeing round corners

Have you ever seen a submerged submarine raise a long tube out of the water so the people on board could see what was going on above the surface? This tube is called a periscope and it uses mirrors to help 'see' round corners. To make your own version you will need two small square mirrors, a piece of stiff cardboard measuring about 1 x 1 foot (30 x 30 cms), some sticky tape, a sharp pencil, a ruler and a knife or pair of scissors.

Step 1

Mark your cardboard into four equal strips and score along the edge of these with a pencil and ruler. Then cut square holes in the second and last strip and two angled slits at both ends of the other two. The slits should be big enough to hold your two mirrors. Be careful to cut your holes and slits in the right positions as shown by the drawing below.

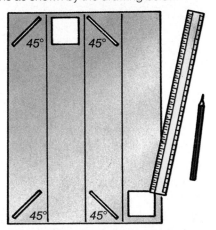

Step 2

Fold the cardboard into a square tube and fix it with sticky tape. Now slide your mirrors into the angled slits. The mirror on the bottom should have its mirrored side uppermost. The one at the top should have its mirrored side angled downwards.

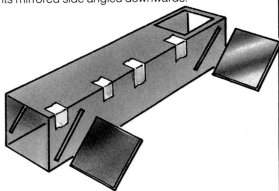

Step 3

Use your periscope to see round corners by looking through the bottom hole. If you sit below the height of a table, you will be able to see objects on top of it reflected in the bottom mirror. This happens because the light from the object you're viewing hits the top mirror, is reflected down to the bottom one and finally into your eyes.

THE WORLD IN FOCUS

Because we look at our world through lenses, it is difficult to imagine what things would look like without them. So in order to find out, let's approach the problem scientifically. First, think about what happens in front of the lens. Remember that we see objects because of the light they reflect. So all the objects in a scene are reflecting light onto the lenses of our eyes. But that light is being reflected at all sorts of different angles and at various strengths, and in various colors. So by the time the light rays reach the front of the lens, they are really just one huge multi-colored blur.

A lens works by refraction. It sorts out the blur, bending the rays from one object in front of the lens so that they come together behind it at one point. This is called the *focus.* It happens for every point in front of the lens, so that you end up with a 'duplicate' scene behind the lens of the picture that's in front of it. If you happen to have some sort of light-sensitive screen handy — like a retina for example — and you put this where the focus is, then you'll be able to 'see' the scene.

Of course, animals have been looking at the world through lenses for millions of years without realizing what a complicated process was going on every time they focused on something. And it wasn't until comparatively recently (about 300 years ago) that we began to understand about lenses and focusing and how they could help us — by correcting bad eyesight, for example, or magnifying things that were too small to be seen with the naked eye. But it took even longer for us to figure out

how to make a piece of equipment that could 'see' the world just like an eye does. And when it was invented everyone thought it was something really new — even though nature had beaten us to it by millions of years! It was, of course, the camera.

It wasn't until the 1800's that the early photographic inventors produced the

world's first photographs by using lenses to create an image on light-sensitive material fixed at the back of a black box. Nowadays, cameras are much more advanced than those early versions and have all sorts of gadgets built into them that can give us a picture of a scene within seconds of pressing a button. In some ways, cameras are now even better than eyes. They can take photographs of things which we can't see – a speeding bullet, for instance, or the wings of a hovering humming bird. They can also make a permanent record of something whereas we have to rely on memories in our 'mind's eye' to recall a scene. And, unlike memories, a photograph will never distort the facts.

31

AUTOMATIC EYE

A basic camera works in much the same way as a human eye. It has an iris like our own eye which controls the camera's aperture (which behaves in much the same way as our pupil) and therefore regulates the amount of light entering the camera. It also has a lens which can be adjusted to focus on either nearby objects or distant scenes. Behind the lens is a small piece of metal called a shutter. This is worked by a spring and normally stops any light from entering the camera. But when you want to take a photograph and press the right button, the shutter opens, the light from the lens passes into the camera and eventually falls onto a piece of photographic film. Just like our retina, the photographic film reacts to the light shining on it so that certain chemical changes take place in its special light-sensitive coating. When the film is finally taken out of the camera and developed, more chemical changes take place so that the image that was focused on the film by the lens becomes visible as a photograph.

In a camera (and in the eye as well), the image that falls on the photographic film is upside-down and back-to-front. This is because the rays from the top part of the scene being photographed pass through the lens and are bent so that they end up on the bottom part of the film. To get everything the right way round again, the film (or negative) has to be transferred onto photographic paper. Luckily for us, our brain puts the picture right for us!

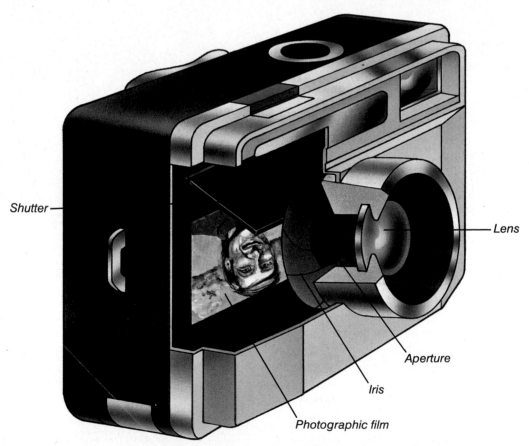

Shutter

Lens

Aperture

Iris

Photographic film

Getting your sight right!

The discovery that lenses could be manufactured and used to focus things was a great breakthrough for people with bad eyesight. Most people who suffer from poor eyesight have one or two problems. *Nearsighted* people have a lens which does not focus the image of an object far enough back to fall on the retina. *Farsighted* people have the opposite problem. Their lens focuses the image too far back, so in effect it falls behind the retina.

Looding through an additional pair of man-made lenses can solve both problems. For nearsightedness people wear spectacles that contain two *concave* lenses. These expand the light beams entering them by just the right amount to focus the image correctly on the retina. For farsightedness, the spectacles are fitted with two *convex* lenses. These converge the light beams to achieve the same result.

Concave lens
expands light beam

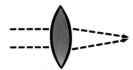

Convex lens
converges light beam

Nearsightedness

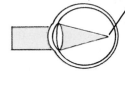

Focus

In this eye, the lens is focusing the image too far forward to fall on the retina

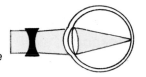

Put a concave lens in front of the eye and it expands the light beam just enough to focus the image correctly.

Farsightedness

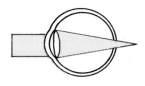

In this eye, the lens is focusing the image too far back to fall on the retina.

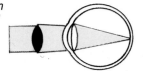

Put a convex lens in front of the eye and it converges the beam just enough to get the focus in the right plane.

Getting the light right

A lot of things can go wrong when you're taking a photograph. You might put your fingers over the lens, or point the camera too low so you get pictures of people with no heads. But even if you don't do any of these things, your pictures can still be ruined if you get the light wrong. They're either nearly all-white because the film has been exposed to too much light, or so dark that you can't see anything.

To help prevent this, some cameras have built-in exposure meters to help you let exactly the right amount of light into the camera. These meters use photocells – electronic devices which respond to light. When light falls on a photocell, it produces a small electric current, and the more light there is, the bigger the current becomes. This current can then be connected up to a meter which measures its strength and therefore the amount of light around.

Photocells can also be used to switch things on and off. Some kinds of street light, for example, are controlled by them – they switch the lights on at night and turn them off in the morning. They can be used as safety devices, too. Many gas central heating boilers have photocells which turn off the gas if the pilot light goes out. And some elevators use them to stop the doors closing if someone is in between them!

IN THE DISTANCE

You can use a single lens as a magnifying glass, to look closely at small objects when you bring them near to the lens. But a single lens won't magnify distant objects. To do that, you need a *telescope,* or a pair of binoculars (which are really two small telescopes fixed side by side).

How telescopes work

The simplest kind of telescope has two lenses, one at each end of a light-tight tube. The largest one (the *objective*) which is farthest away from your eye, produces an image of the object you are viewing, and

the smaller lens (the eyepiece) magnifies this image to give you a close-up view.

The trouble with this kind of telescope is that it produces an upside-down image, so telescopes for general use have extra lenses in between the objective and the eyepiece to turn the image the right way up. Another way of doing this, which is used in good quality binoculars, is to use a pair of prisms between the lenses to reflect the image the right way up. Telescopes like these which use a combination of lenses or lenses and prisms are known as refracting telescopes, or *refractors* for short.

Here the open dome of the Royal Observatory at Greenwich, London, reveals the end of a refracting telescope.

Reflecting telescope

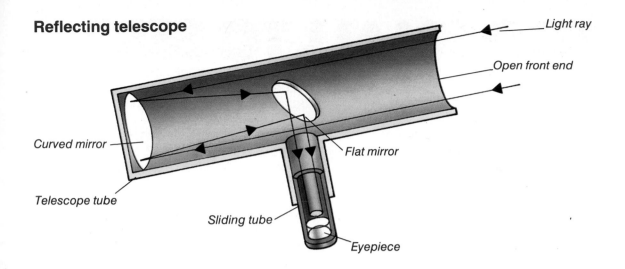

Light ray

Open front end

Flat mirror

Curved mirror

Telescope tube

Sliding tube

Eyepiece

Another type of telescope, the *reflector,* uses a curved mirror instead of an objective lens. With this, the front end of the telescope tube is open, and the light passes down it to the mirror, which is at the other end. The curved mirror focuses its reflected image on to a second small, flat mirror which finally directs the image into the eyepiece which is in the side of the telescope tube. Most of the telescopes used by astronomers are of this sort because the mirror doesn't distory the image or create color fringing in the way that a lens can.

Big apertures, bright images

The diameter of the objective lens or mirror is called the *aperture* of the telescope. The bigger it is, the brighter the image it makes. To astronomers, a bright image is more important than a very high magnification, since it enables them to see the very faint light from far-off stars.

It is not practical to make a lens with a diameter of more than about 1 yard (about 1 meter) any bigger than this and it would sag under its own weight. But mirrors can be made much bigger, so the really powerful astronomical telescopes all use mirrors instead of objective lenses. The biggest in the world, at Zelenchukskaya in the USSR, has an aperture of 20 feet (6 meters).

Science discovery

Telescopes and astronomy

The first telescopes were produced around the beginning of the 17th century, and a few years later the Italian scientists Galileo Galilei had the idea of using one to study the stars. By doing this, he began the era of telescope astronomy, but he also got himself into a lot of trouble!

A century earlier, the Polish astronomer Nicolas Copernicus had suggested that the Earth and the other planets of our universe revolve around the Sun. This was as shocking suggestion at the time, because people then thought that everything revolved around the Earth!

So when Galileo said that his observations with the telescope showed that Copernicus was right, he was in big trouble. Luckily, his work had already been published and was being studied by scientists all over Europe. Before too long, people were beginning to accept the idea that the Sun, not the Earth, was the center of the solar system.

TOO SMALL TO SEE.

No matter how powerful a telescope is, you can't use it like a sort of extra-powerful magnifying glass to look at very small objects. It can't be focused on things that are very close to it. So for really powerful close-up magnification, you need a *microscope*.

Like a telescope, a basic microscope has an objective lens and an eyepiece. The objective lens produces an enlarged image of objects very close to it. And this is magnified further by the eyepiece lens.

To use a microscope, you first put the specimen you want to examine on to a glass slide. Then you put the slide on to a platform beneath the objective lens. On all but the simplest microscopes, there is an electric lamp which is used to light up the specimen, sometimes by being reflected by a mirror. If the specimen is transparent, the light is shone up through it from below.

Modern laboratory microscopes usually have several objectives of different magnifications. These are mounted on a nosepiece, which can be rotated to bring the objective you want into position above the specimen slide. Many also have two eyepieces so you can view your specimen using both eyes, rather like looking through binoculars. The light from the objective is diverted into the eyepiece by prisms.

The amount of magnification a light microscope gives can be from about 30x actual size up to 200x. It is possible to make light microscopes even more

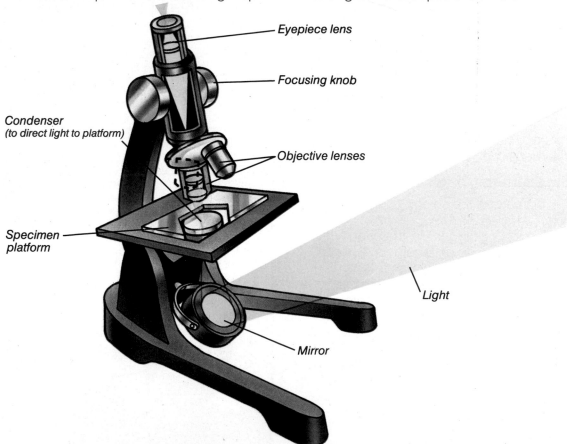

Eyepiece lens

Focusing knob

Condenser
(to direct light to platform)

Objective lenses

Specimen platform

Light

Mirror

powerful than this, but it would be a waste of time. This is because if a light wave meets anything much smaller than its own wavelength (about 0.4 millionth of a yard) it will just curve around it and carry on as if it wasn't there! The light won't be affected by the object – by being reflected or refracted – so, in theory, the object is 'invisible'!

The only way you can get to see details of objects smaller than this is to use an *electron* microscope. This instrument uses rays of electrons rather than light. Like the photons that make up light, electrons can travel in waves. But the wavelength of electrons is much smaller than that of light, so they can capture much finer detail.

Seen through a microscope, the fantastic eyes of a fly are revealed. Unlike our eyes, those of many insects are made up of numerous light-sensitive units.

Science discovery

The first microscope

When Galileo (see page 35) was experimenting with telescopes, he realized that lenses could also be used for magnifying very small objects. And by the middle of the 17th century, many eminent scientists were experimenting with microscopes.

One of the most successful of these was a Dutch draper and spare-time biologist called Anton van Leeuwenhoek. His microscopes consisted simply of small, single lenses mounted in brass holders. But he was so skilled at making lenses (he made over 400 in all) that they could magnify by up to almost 200 times. He studied all sorts of biological speciments, and even reported seeing what were probably bacteria, although he himself had no idea what they were at the time.

Microscopes are great for looking at tiny things — like small insects, plant parts and so on. You can make your own very simple model by following the instructions here.

Experiment 5

Make a microscope

To make your own microscope you will need a strip of metal, a glass, a small mirror, some clay, some sticky tape and the use of a drill.

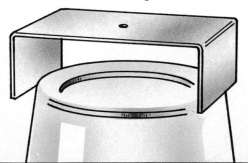

Step 1

Ask your parents to help you drill a small hole, about $\frac{1}{20}$ inch (1 mm) in diameter, in the middle of your metal strip. Then bend the strip as shown so that its top is just wider than the diameter of the glass's bottom.

Step 2

Turn the glass upside down and tape the metal strip to it as shown. There should be about $\frac{3}{8}$ inch (1 cm) between the glass and the strip.

Step 3

Prop your mirror up on some clay so that it reflects light upwards. Then place the glass over the top.

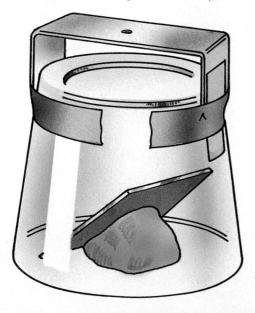

Step 4

Carefully drop some water into the hole in the strip. This will act like a lens to enlarge any object that you place beneath it, resting on the glass.

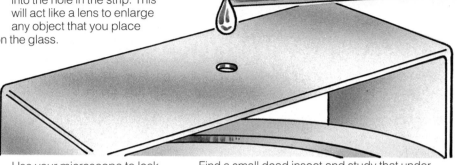

Step 5

Use your microscope to look at things in detail. Pull a hair from your head and look at it. Can you see the white blob at the end? This is the only part of the hair that's alive, full of lots of cells that are multiplying to add to the rest of the hair's shaft.

Find a small dead insect and study that under your 'scope. Can you see all the different parts of its body? Does it have compound eyes like the fly on page 37? You could also try looking at a bird's feather. Can you see the tiny hooks on the edge of each barb? These catch onto each other to mesh the feather together.

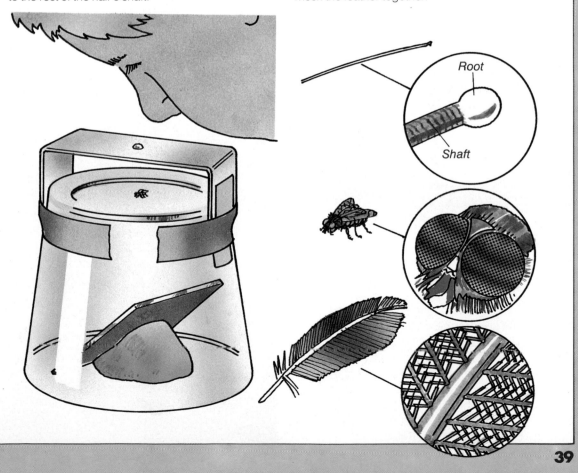

Root

Shaft

WHY DOES...?

...A rainbow sometimes appear in the sky?

Every so often a shower of rain will be lit up by the Sun and a rainbow will appear in the sky. This happens because when the Sun's rays hit the raindrops they are both reflected and refracted by them. The white light enters the droplets and is reflected back out of them and into your eyes. At the same time, the water acts like a glass prism to split the light up into its different colors. These colored lights then come together to form a semicircle of color in the sky.

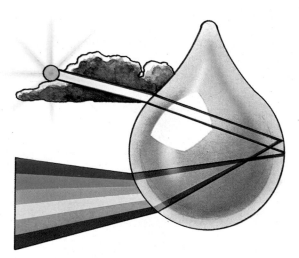

...Your skin turn brown in the sun?

The color of your skin depends on the amount of a substance called *melanin* within it. The invisible ultraviolet light from the Sun can penetrate your skin and damage the tissue beneath it. So when light skin is exposed to strong sunlight, it produces more melanin, which helps to protect the skin and the flesh underneath from the harmful effects of the ultraviolet light.

...The sound of thunder never accompanying the flash of lightning?

You've probably noticed that in a thunderstorm, you see the flash of lightning before you hear the sound of thunder. In fact the two actually happen at the same time. The reason for the delay between them is that light travels so much faster than sound. The light from the lightning flash reaches you almost instantly, but the noise of the thunder takes a while longer.

You can calculate how far away the lightning storm is by counting the number of seconds between the lightning and the thunder. For every 3-second delay, the distance is about $6/10$ mi (1 km). If you only hear the thunder, the lightning's missed you!

Jagged streaks of lightning illuminate the sky seconds before the thunderclap reaches our ears

... Polarized glass make the best sunglasses?

Ordinary sunglasses use tinted glass (or plastic), which simply cuts down the total amount of light reaching your eyes. Polarized lenses also do this, but they cut down the amount of 'glare' as well – the light reflected off the surface of sunlit water, for instance, which stops you being able to see into the water. The waves of sunlight normally vibrate up and down, but a lot of the sunlight reflecting off horizontal surfaces, such as water, vibrates from side to side as well. This is what causes glare. Polarized sunglasses only let the up and down waves pass through, so the glare of the side to side waves is filtered out.

– thanks to the fantastic speed at which light travels.

...The sky look red at sunset?

When the Sun is low in the sky, its light is passing through the atmosphere of the Earth at a very low angle. As it sets, or rises, its light is traveling almost horizontally to reach us, so it has to travel through a lot more air than it does when it is shining down from high in the sky at midday. The molecules of air through which it passes scatter the light, but the longest wavelengths (which are the red ones) get through with less scattering than the other colors. This is why the Sun, and the sky around it, often looks red at sunset or sunrise. Sometimes the Sun is surrounded by other colors, such as orange and yellow, which also have fairly long wavelengths. Try shining a flashlight through a jam jar that contains water mixed up with a bit of milk. The liquid will take on a sort of pinkish color because, again, most of the wavelengths are being scattered by the milk molecules so only the red gets through.

...A star twinkle?

Have you ever noticed the 'heat' rising from the ground on a hot day? This shimmering effect, which sometimes makes it look as though there is water on the ground, is caused by the swirling of the heated air as it rises from the hot ground. Rays of light passing through this swirling heat will become twisted and distorted, creating an effect that makes people think they can see water in a waterless desert. And at night, warm air from the ground rising through the colder upper air has a similar effect – making the stars appear to twinkle, although their light is really as steady as that of the Sun.

THINGS TO REMEMBER

Here are some explanations of some of the words in this book that you may find unfamiliar. In some cases, they aren't the exact scientific definitions, because many of these are extremely complicated. But the descriptions should help you to understand what the words mean.

ANGLE OF INCIDENCE The angle at which light hits a mirror, measured from a line at right angles to the mirror's surface.

ANGLE OF REFLECTION The angle at which light bounces off a mirror, measured from a line at right angles to the mirror's surface. The angle of reflection and the angle of incidence are always equal.

CONVEX LENS A lens which converges light beams.

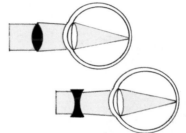

CONCAVE LENS A lens which expands light beams.

ELECTROMAGNETIC SPECTRUM Types of energy that travel like waves and include radio waves, light, x-rays and gamma rays.

EYEPIECE The small lens of a telescope or microscope nearest to the eye.

FREQUENCY The number of waves per second leaving a source.

INFRA-RED ('Below the red') Light with a wavelength just longer than the red end of the visible spectrum. Infra-red light is invisible to our eyes but we feel it as heat

IRIS Part of an eye or camera which controls the amount of light entering by altering the size of the pupil or aperture.

LASER Special device which produces a concentrated beam of light. Laser stands for **L**ight **A**mplification by **S**timulated **E**mission of **R**adiation.

LENS A specially-shaped piece of glass which uses refraction and can be used to focus images and make them bigger or smaller.

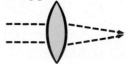

LIGHT   Electromagnetic energy with a wavelength of 0.38-0.75 microns (millionths of a meter), to which our eyes react.

LIGHT YEAR   The distance light travels in one year – 5.88 million million miles (9.46 million million kms).

LUMEN   Unit used to measure the amount of light falling on a surface or passing through a place, as lumens per square foot or meter.

MELANIN   Chemical substance in the skin that reacts to ultraviolet light to help protect us from its harmful effects.

MICROSCOPE   A device used to magnify small objects at close range.

OBJECTIVE LENS   The largest lens in a telescope which focuses a distant image.

PHOTOCELL   Electronic device which responds to light by giving off an electric current.

PHOTONS   Particles of light energy.

PHOTOSYNTHESIS Chemicals process in plants which is powered by light energy.

PRISM   A solid, flat-sided piece of glass used for refracting light.

PUPIL   A circular opening at the front of the eye which lets in light.

REFLECT   To bounce off.

REFLECTOR   A type of telescope which uses a curved mirror instead of an objective lens.

REFRACT   To bend.

REFRACTOR   A type of

telescope which uses prisms as well as lenses to focus an image.

RETINA   Light-sensitive lining at the back of the eye.

TAPETUM   Membrane behind a cat's eye which reflects light.

TELESCOPE   A device used to enlarge the image of a distant object.

ULTRAVIOLET   ('Beyond the violet') Light with a wavelength just shorter than the violet end of the visible spectrum. UV light is invisible to our eyes but some creatures, such as bees, can see it.

WAVELENGTH   The distance between the peak of one light wave and the peak of the next.